SANTA ANA PUBLIC LIBRARY

D0700084

Understanding the Elements of the Periodic Table™

THE BORON ELEMENTS

Boron, Aluminum, Gallium, Indium, Thallium

Heather Hasan

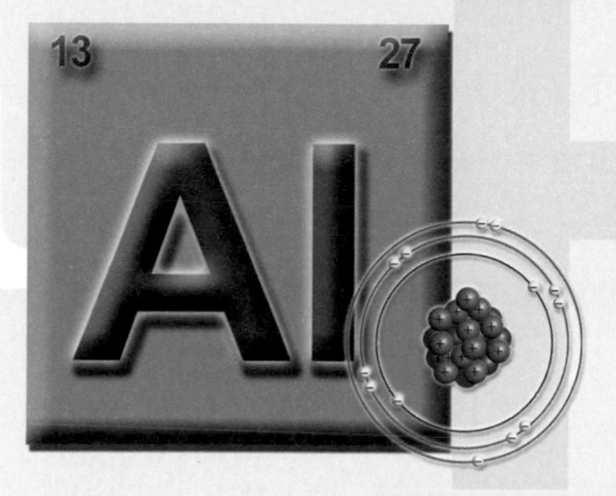

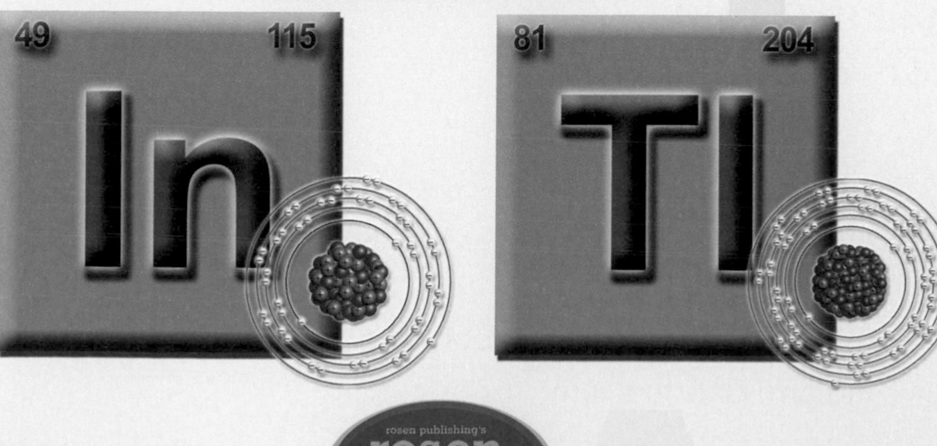

New York

To my daughter, Sarah Elizabeth—you are a blessing from God

Published in 2010 by The Rosen Publishing Group, Inc.
29 East 21st Street, New York, NY 10010

First Edition

Library of Congress Cataloging-in-Publication Data

Hasan, Heather.
The boron elements: boron, aluminum, gallium, indium, thallium / Heather Hasan.—1st ed.
p. cm.—(Understanding the elements of the periodic table)
Includes bibliographical references and index.
ISBN-13: 978-1-4358-5333-1 (library binding)
1. Boron—Juvenile literature. 2. Group 13 elements—Juvenile literature.
3. Periodic law—Juvenile literature. I. Title.
QD181.B1H37 2010
546'.67—dc22

2008049315

Manufactured in the United States of America

On the cover: The atomic symbols and structures of boron, aluminum, gallium, indium, and thallium.

Contents

Introduction

The ancient Greeks believed that there were four basic elements: air, earth, fire, and water. Today, we know that there are more than one hundred different elements.

Everything around us is made up of these elements. No one really thought to organize these elements until the middle of the nineteenth century, when a Russian scientist named Dmitry Mendeleyev published the first periodic table in 1869. Though the periodic table we use today is a little different from Mendeleyev's, we owe him a great deal of gratitude.

The periodic table is as important to chemists as maps are to explorers. The periodic table arranges the elements according to their periodic properties. This helps chemists to understand and predict how the different elements act.

There are eighteen groups (columns) on the periodic table. The elements in a given group have similar characteristics. Chemists can tell a lot about an element just by knowing what group it is in. In this book, we will examine the elements in group 13, also known as the boron group. Of the group 13 elements, boron and aluminum are the most well-known. This book will also focus on the lesser-known group 13 elements (gallium, indium, and thallium). We will discuss where they are found, what makes them unique, what their properties are, how they react, and how they are used.

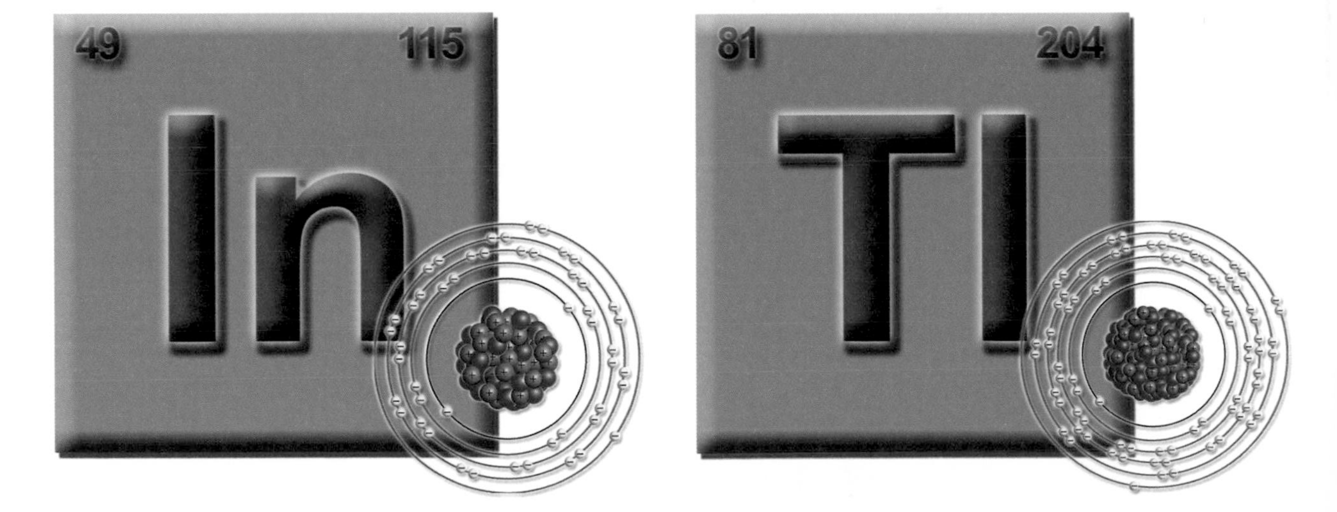

The group 13 elements include aluminum (Al), gallium (Ga), boron (B), indium (In), and thallium (Tl). This group is also called the boron group.

Chapter One
Group 13

The boron group is the series of elements in group 13 of the periodic table. The boron group consists of boron (B), aluminum (Al), gallium (Ga), indium (In), and thallium (Tl). The group 13 elements are found in the ground. In fact, aluminum is the third most abundant element in Earth's crust, following oxygen (O) and silicon (Si). The other group 13 elements are far less abundant. Indium, for example, is only the sixty-first most abundant element in Earth's crust, according to the U.S. Ecological Survey.

The group 13 elements are present in various ores and minerals within Earth's crust. These ores and minerals are mined from the earth so that people can make use of them. Aluminum is by far the most widely used element from this group. We use aluminum in many ways in the home, in industry, and in transport. We use it to make such things as window frames, power cables, aircraft parts, cooking foil, and drink cans. Boron is used to make semiconductors, glass, and cleaning products. It is also used in such things as pesticides, fire-proof fabrics, and cancer treatments. Gallium is used in semiconductors, supercomputers, and mobile phones. Thallium is used for making such things as special low-melting-point glass, photoelectric cells, and nuclear medicine. Indium does not have many uses, but we use it to make things like sprinkler systems and semiconductors.

Earth's crust is the thin, outermost layer of the planet. The ores from which the group 13 elements are extracted are found in the crust.

Where Are the Boron Elements?

The group 13 elements are present in various ores. An ore is a mineral deposit from which a metal can be extracted. The main ores from which boron is extracted are kernite, tincal (borax), ulexite, and colemanite. All of these ores are called borates, which are compounds that contain oxygen and boron, along with other elements.

Most of the aluminum in the world today is made from bauxite. Bauxite contains aluminum combined with oxygen and hydrogen (H), along with various other elements. Though gallium, indium, and thallium are also all present in ores, no ores are mined specifically for their production. These metals are all obtained as by-products in the production of other metals.

Gallium is produced as a by-product of aluminum and zinc (Zn) production. Indium is extracted as a by-product of zinc smelting. Thallium is obtained as a by-product of the smelting of zinc, copper (Cu), and lead (Pb) ores.

The History of the Boron Elements

The use of boron can be traced back to the ancient Egyptians. They used a compound of boron, known as borax, to join, or solder, metals

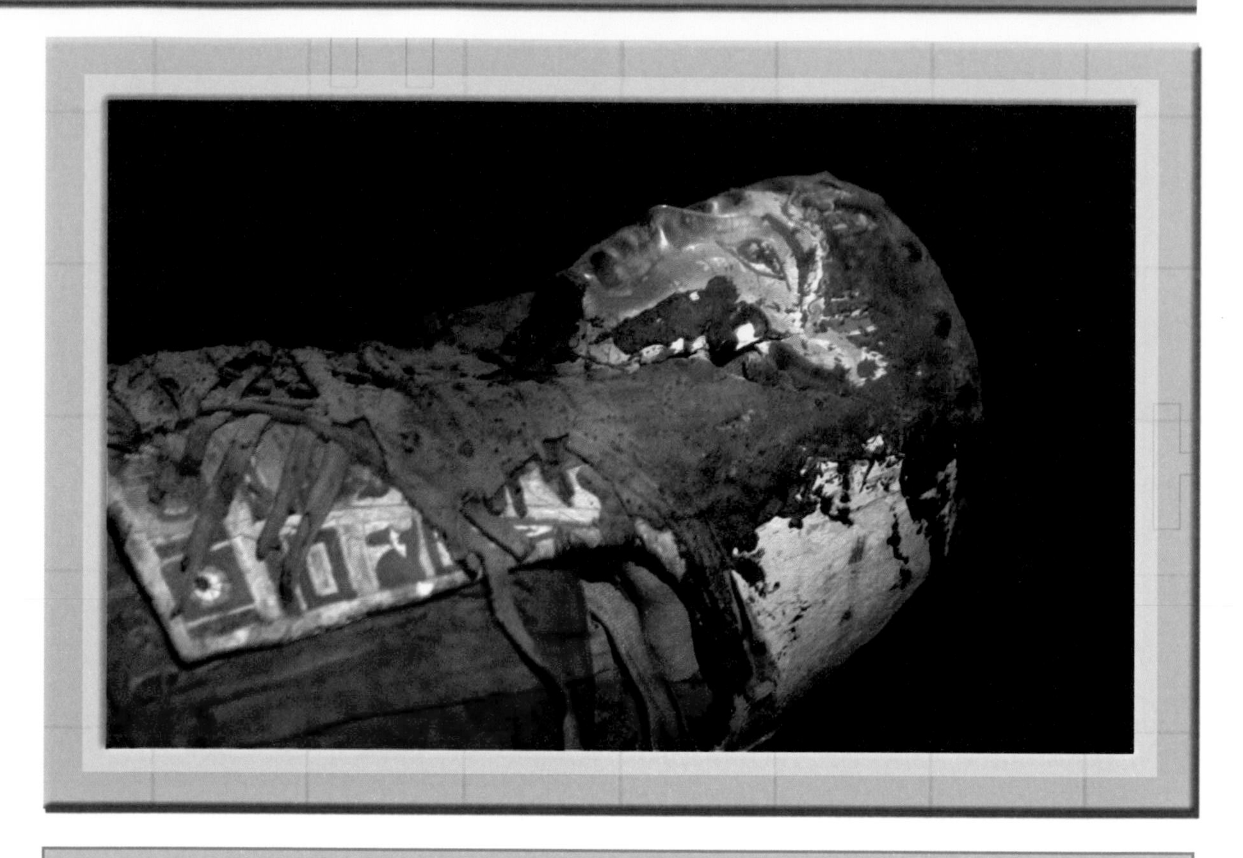

Borax, a boron compound, was used by the ancient Egyptians to preserve mummies, such as this 2,000-year-old specimen.

together. They also used it in medicine and for mummification. The ancient Egyptians, the Greeks, and the Romans all used a compound of aluminum, known as alum, for dyeing fabrics.

The other elements of the boron group were first discovered in the late nineteenth century. They were detected with a spectroscope, an instrument used to analyze the light given off by elements when they are heated. Every element emits different light when it glows. The spectrum of an element consists of a series of colored lines. When scientists look into a spectroscope, they can tell what elements are in a substance by what color lines they see.

Thallium was discovered in 1861 by Sir William Crookes (1832–1919) of England. Crookes was looking at some impure sulfuric

Sir William Crookes, shown here in 1850, discovered thallium when he observed the green light it emitted under a spectroscope.

acid in a spectroscope. The impurity in the sulfuric acid was thallium. Crookes was the first to see the green light that thallium gives off when it is observed in a spectroscope. Crookes named this unknown element thallium because the green line that he observed reminded him of a fresh green shoot (*thallos* is a Greek word that means "a green shoot or twig").

Crookes reluctantly shared the credit for discovering thallium with a physicist from France named Claude-Auguste Lamy (1820–1878). Lamy was the first to separate the metal out, making a small nugget of it. Lamy sent his small piece of metal to the London International Exhibition in 1862, where it was proclaimed a new element. For his achievements, Lamy was awarded a medal. This made Crookes furious, for he felt that he was the true discoverer of thallium. Crookes campaigned to have the award given to him, and eventually, he was also given a medal for the discovery of thallium.

In the year following its discovery, thallium was found to be present in many places, including wine, tobacco, spring water, and sugar beet. In 1866, a thallium mineral was discovered in Sweden. It was named crookesite in honor of William Crookes's discovery. In all of these cases, thallium's green glow allowed it to be easily detected.

Indium was discovered in 1863 by Ferdinand Reich (1799–1882) and Hieronymus Richter (1824–1898) at the Freiberg School of Mines in Germany. Reich was a professor of physics, and Richter was a professor of chemistry. Reich was examining a sample of zinc ore that he thought contained the recently discovered thallium. Reich was color-blind, so he called in his colleague, Richter, to look at the spectrum. Richter immediately saw a brilliant violet line. They named the new element indium after the indigo (violet) line. The men isolated a sample of the metal and together published a paper announcing its discovery. People began searching for minerals of the new element. However, nothing was found until 1876, when indium-zinc ores were discovered in Colorado and in Bergamo, Italy.

Gallium was discovered by Paul-Émile Lecoq de Boisbaudran in France in 1875 as he examined a zinc ore in a spectroscope. In a spectroscope, the light that gallium gives off shows up as two violet lines. Boisbaudran had never seen these lines before, so he knew an unknown element was present in the blend. He named the element gallium, a name derived from the Latin word *gallia*, which means "France." Boisbaudran was able to isolate a sample of the metal, and in 1875, he presented it to the French Academy of Science. However, the existence of gallium had already been predicted six years prior to this by the Russian chemist Dmitry Mendeleyev, when he developed the periodic table.

The Periodic Table

The periodic table that we use today is based on the table that Dmitry Mendeleyev published in 1869. Mendeleyev developed his version of the periodic table while teaching chemistry at the University of St. Petersburg in Russia. He wanted to organize the elements in a way that would make it easier for his students to study and understand them. At that time, there were only sixty-three known elements. Mendeleyev wrote the name of each element on a card along with everything that was known about each of them. He then looked for a logical way to organize them.

Mendeleyev decided to arrange the elements in horizontal rows, according to atomic weight, with the lightest element of each row on the left end and the heaviest on the right. He started a new row whenever he found an element similar to the first one. He noticed that when he arranged the elements like this, the elements in each column behaved in similar ways.

Though Mendeleyev's periodic table did not list all of the elements that we know of today, boron, aluminum, indium, and thallium were among those that he included on his first chart. Though gallium had not been discovered yet, Mendeleyev saw that there should be an element like it, so he left a space for it in his table. He predicted the existence and properties of this element based on his periodic table. The accuracy of his description of gallium helped to convince scientists of the soundness of his table.

Chapter Two
A Look at the Elements

Everything on Earth, as well as throughout the universe, is made up of one or more elements. Each of these elements is made up of a different type of atom. An atom is the smallest unit of an element. The atoms of aluminum are different from those of thallium. Because they all have different atoms, each element is unique. Each has a different look and feel, and each behaves differently. However, every atom of a given element, such as indium, is exactly the same. Atoms are very tiny, but amazingly, atoms are made up of even smaller components called subatomic particles. In order to truly understand what makes elements unique, we have to take a closer look at these particles.

Subatomic Particles

There are three subatomic particles that make up the atom: neutrons, protons, and electrons. Neutrons and protons are clustered together at the center of an atom to form a dense core called the nucleus. Neutrons carry no electrical charge, while each proton has a positive electrical charge. This gives the nucleus an overall positive electrical charge. For instance, a gallium atom has thirty-one protons in its nucleus, so its nucleus has a charge of +31. Going down a column in the periodic table, the number of protons in each atom increases. Boron, at the top of

The atomic structure of boron is composed of five positively charged protons in the nucleus surrounded by five negatively charged electrons, arranged in shells.

the column, has five protons in an atom, and thallium, at the bottom of the column, has eighty-one. The number of protons in an atom of an element determines the element's atomic number. This number is often found to the upper left of the element's chemical symbol in the periodic table.

Electrons are negatively charged particles that are arranged in layers, or shells, around the nucleus of an atom. The electrons are not fixed in a single position but circulate around the nucleus. In a neutral atom, the number of protons and electrons is equal, so the positive and negative charges of the atom balance out. For example, since indium has forty-nine protons, it also has forty-nine electrons.

Groups and Periods

Going across the periodic table from left to right, each horizontal row of elements is called a period. Elements in the same period have the same number of electron shells around the nucleus of their atoms. Boron is in period 2, so it has two shells of electrons surrounding its nucleus. Aluminum (period 3) has three, gallium (period 4) has four, indium (period 5) has five, and thallium (period 6) has six shells of electrons surrounding its nucleus. The electrons in the outmost shell are called valence electrons. It is these electrons that determine how an element acts.

The Poisoner's Poison

Thallium has been nicknamed "the poisoner's poison" and "inheritance powder." Among the symptoms of thallium poisoning are weakness, nausea, pain in the arms and legs, and loss of hair. Thallium's ability to make people lose their hair led to its initial use as a depilatory (hair remover). However, once its toxicity was discovered, this use ceased. Thallium was once an effective murder weapon because it is odorless and tasteless. Agatha Christie even used thallium as a murder weapon in her novel *The Pale Horse*. Though cases of thallium poisoning still occur, doctors now know what symptoms to look for. An antidote (prussian blue) has also been discovered.

As you read down the periodic table from top to bottom, each vertical column of elements is called a group. All of the elements in a group have the same number of electrons in their outermost electron shell. Because of this, the elements in a group have some similar properties. The elements of group 13 (boron, aluminum, gallium, indium, and thallium) each have three electrons in their outermost electron shells. This gives them some similar properties, but each one is unique and has its own different properties, too.

The Boron Elements

Unlike Mendeleyev's chart, the periodic table that we use today lists the elements in order of increasing atomic number (the number of protons). When the elements are arranged like this, many patterns in their properties can be seen. By seeing where an element is located on the periodic

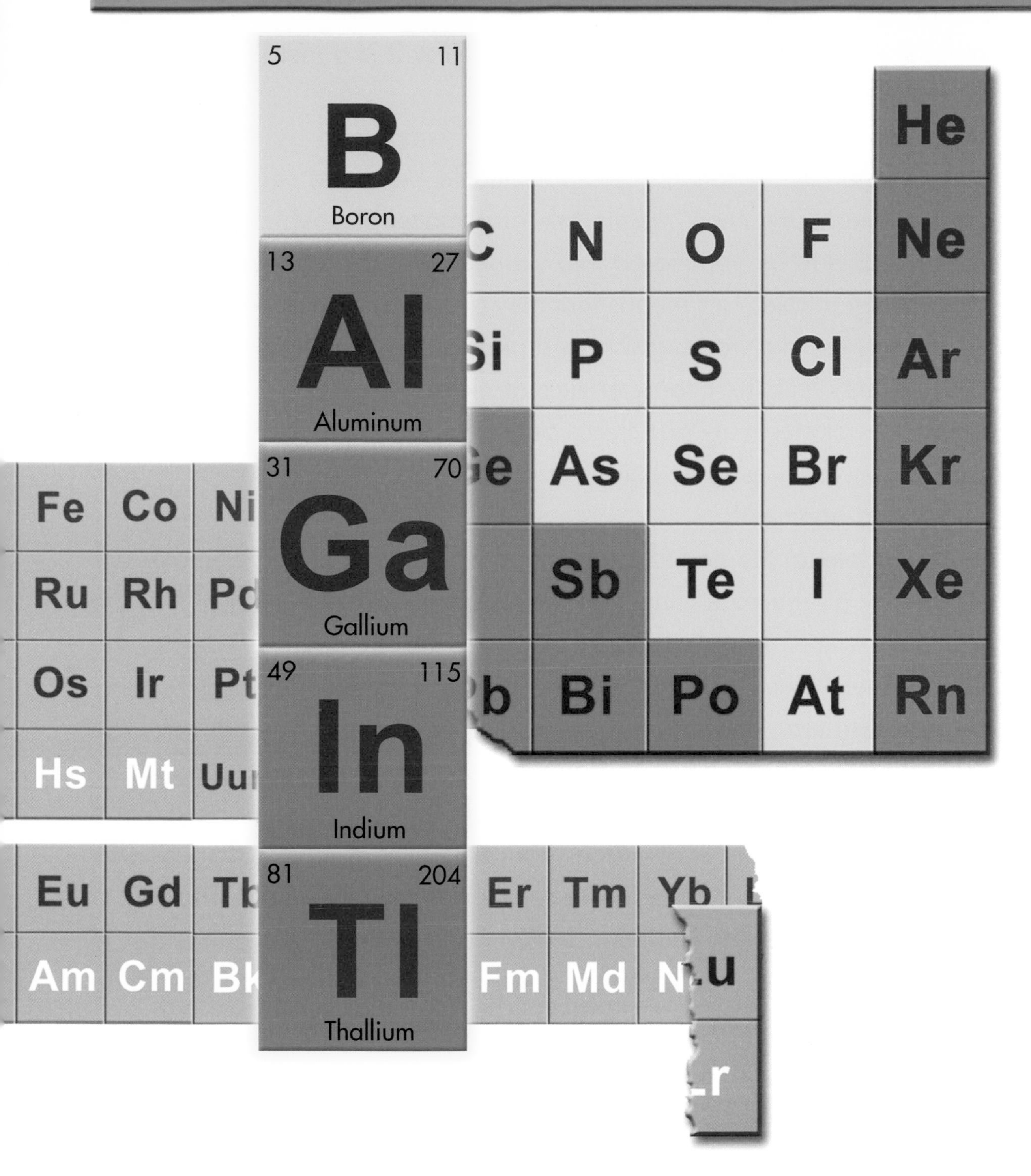

There are many trends on the periodic table that help us classify the elements. For example, as we move down the group 13 column, the elements become more metallic in nature.

table, scientists and you can figure out whether it is a metal, a nonmetal, or a metalloid.

Metals, such as aluminum, are easily recognized by their physical traits. Generally, metals can be polished to be made shiny. They also conduct electricity. Most metals are also malleable, which means that they are able to be hammered into shapes without breaking. Metals are also usually ductile. This means that they are able to be pulled into wires.

Substances like wood, glass, and plastics are classified as nonmetals because they lack the characteristics of metals. They are generally poor conductors of electricity and heat, are brittle, and are not easily worked into shapes or wires. Metalloids, or semimetals, have properties of both metals and nonmetals.

If you look at the periodic table, you will notice that the elements are divided by a "staircase" line. The metals are found to the left of this line and the nonmetals on the right. Most of the elements bordering the line are metalloids. Though aluminum is one of the elements that borders the staircase line, it is the one exception. Aluminum does not have the characteristics of a metalloid but rather those of a metal. Boron, on the other hand, is classified as a metalloid. Though it is very hard like a metal, it breaks easily. It also does not conduct heat or electricity as well as other metals, but it conducts better than other nonmetals. The rest of the group 13 elements (thallium, indium, and gallium) are all metals. In general, the group 13 elements clearly show the trend of becoming more metallic in nature when moving down the column.

Chapter Three
The Properties of the Boron Elements

The properties of an element are the characteristics that are used to identify or describe it. An element's properties can be divided into two categories: physical and chemical. The physical properties of any material are those that can be seen without combining it with some other substance. Some examples of physical properties are a material's appearance, its phase at room temperature (whether it is a solid, liquid, or gas), its density, and its hardness.

The chemical properties of a material describe the material's behavior when it is combined with other substances. These properties can be observed only during a chemical reaction. During a chemical reaction, substances combine and change into different substances. If a substance easily undergoes chemical change with another substance, it is said to be very reactive.

Boron has several forms. Its most common form is a dark gray powder. Pure aluminum is a silvery white metal. Very pure gallium has a beautiful, silvery appearance. Indium is also a silvery white metal with a brilliant luster. A fresh surface of thallium displays a metallic luster, but in air, it soon turns a bluish gray color, resembling lead in appearance.

At room temperature, an element is found in one of three phases: solid, liquid, or gas. Knowing the phase, or physical state, of an element at room temperature helps scientists to identify it. All of the boron

Boron is most commonly found as a dark gray powder. Knowing the physical properties of a substance, such as its appearance, can help you to identify it.

elements are found in the solid phase at room temperature. A solid has a fixed shape and volume. Solids also resist being compressed, or having their shape changed.

The Phase Changes of the Boron Elements

If solid substances are exposed to high enough temperatures, they will turn to liquid. The temperature at which this phase change occurs is called the melting point of the substance. If heated even further, a liquid will boil and become a gas. The temperature at which this phase change occurs is called the boiling point. For a substance to change phases, the forces holding its particles together must be overcome. Metals often have high melting and boiling points due to the strong bonds that hold their atoms together. Even so, there is a considerable variability in the melting and boiling points of metals. Mercury (Hg), for example, melts at -38 degrees Fahrenheit (-39 degrees Celsius), while tungsten (W) melts at 6,170°F (3,410°C), the highest melting point of any element.

The melting and boiling points also differ within group 13. Boron, the only metalloid of the group, has a very high melting point of 4,172°F (2,300°C). Boron's boiling point is also high, 6,614°F (3,658°C). The other four group 13 elements differ from boron in that they are low-melting metals. Aluminum's melting point is 1,221°F (661°C), and its boiling point is 4,473°F (2,467°C). Gallium's melting point is 86°F (30°C). It is one of only a few elements that occur as liquids at or close to room temperature. Gallium's melting point is so low that it can be melted by body temperature (about 98.6°F [37°C]). It will turn to liquid when it is simply held in the palm of a person's hand. Gallium's boiling point is about 4,400°F (2,400°C).

With a melting point of 86°F (30°C) and a boiling point of 4,352°F (2,400°C), gallium has a liquid range of 4,266°F (2,370°C). This is one of the largest liquid ranges of any metal. It is for this reason that gallium is used in high-temperature thermometers. When gallium does solidify, it expands. It is one of only a few substances that does this. For this reason, it should not be stored in glass or metal containers. These types of containers could break as the gallium solidifies.

Gallium also has the unusual property of being able to be supercooled below its melting point. Supercooling involves cooling a substance below its melting point without it becoming a solid. Since gallium's melting point is 86°F (30°C), one would expect it to be a solid at 85.46°F (29.7°C). However, it is fairly easy to cool gallium below 85.46°F (29.7°C) without it solidifying.

Indium's melting point is 313°F (156°C), and its boiling point is 3,767°F (2,075°C). Indium has the unusual property of remaining soft

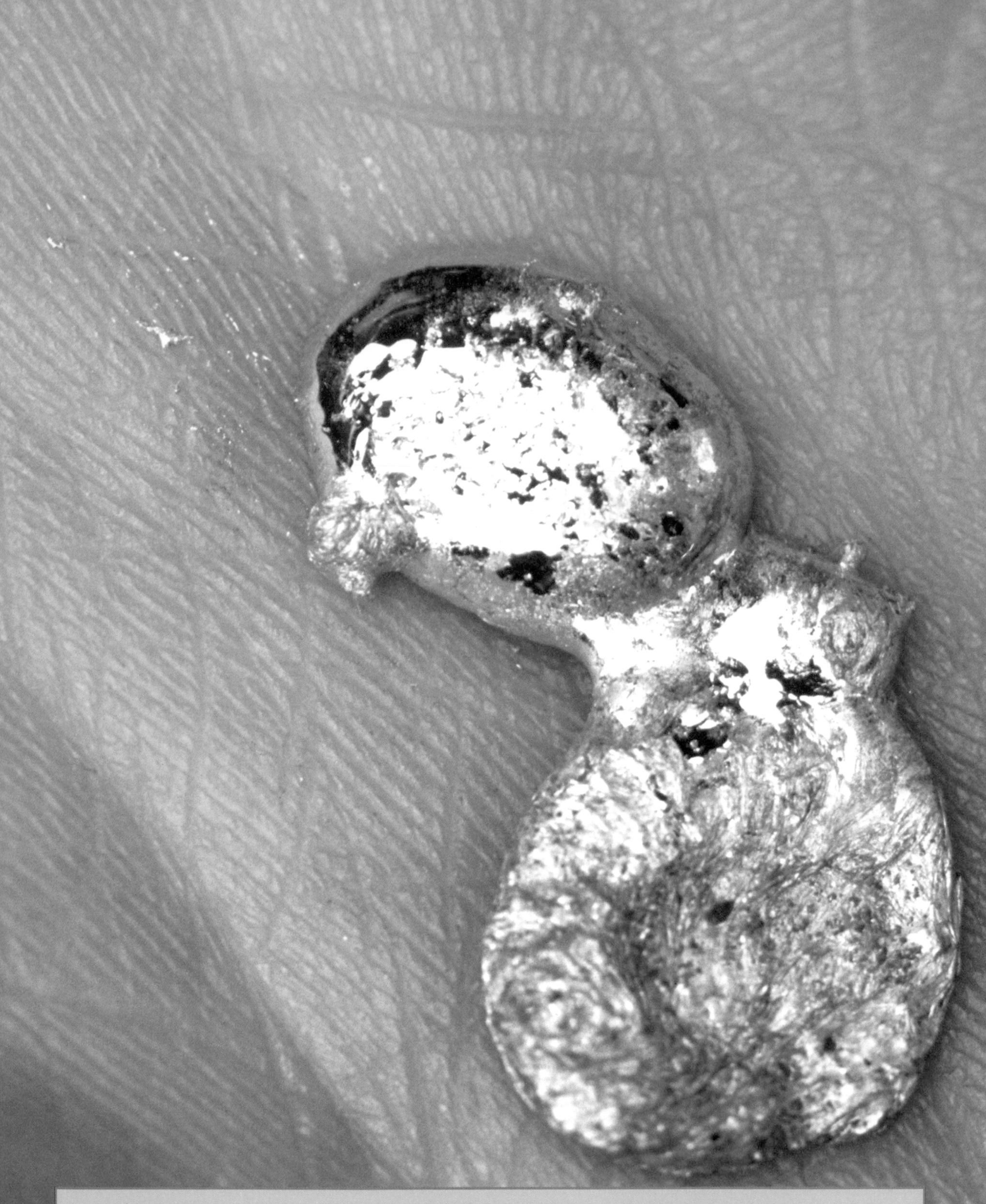

Gallium, with a melting point of 86°F (30°C), is shown melting in the palm of someone's hand. The temperature of the human body is about 98.6°F (37°C).

and workable even at low temperatures. For this reason, we use it in special equipment that is used at temperatures near absolute zero. Absolute zero is the coldest temperature possible, about -459°F (-273°C). Thallium also melts easily. Thallium has a melting point of 579°F (304°C) and a boiling point of 2,655°F (1,457°C).

The Sighing Metal

An unusual physical property of indium occurs when the metal is bent. When indium is bent (when it is stressed), it produces a sigh-like sound. This is called a "tin cry" because it is also characteristic of the more well-known metal tin (Sn). A bar of indium will "cry" repeatedly when it is bent. This is yet another property that can help identify the metal.

Density

Density is another physical property of matter. Each element has a unique density associated with it. Density refers to the mass contained within a unit volume. It describes how compact an object is. The densities of the group 13 elements are listed in the following table. The densities get larger down the group. Solids generally have slightly higher densities than liquids, which, in turn, have much higher densities than gases.

Element	Density in g/cm^3
boron	2.34
aluminum	2.70
gallium	5.90
indium	7.30
thallium	11.85

In chemistry, the densities of many substances are compared to the density of water (1.0 g/cm^3). If an object with a lower density than water (H_2O) is placed in water, the object will float. However, if an object has a higher density than water, it will sink. All of the boron elements have higher densities than water. Therefore, they would all sink when placed in water.

Hardness

The group 13 elements are mostly soft metals. Once again, boron (the only metalloid in the group) is the exception. Boron has a hardness of

In mineralogy, hardness is a measure of the "scratchability" of a material. On the Mohs hardness scale, materials with higher numbers are harder and will scratch those materials with lower numbers.

9.3 on the Mohs hardness scale. The Mohs hardness scale measures the hardness of minerals, metals, and other solids. The scale ranges from 1 to 10. A material with a hardness close to 10, such as boron, is very hard. For comparison, a diamond has a hardness of 10. A material with a hardness near 1 is quite soft. On the Mohs hardness scale, aluminum has a hardness of 2.75.

Gallium, indium, and thallium are even softer than aluminum. The hardness of these elements are 1.5, 1.2, and 1.2, respectively. All three of these metals are so soft that they are able to be cut with a knife. Indium and thallium are two of the softest metals known. If they are rubbed across a piece of paper, they will leave a trail of metal particles on the paper, forming a mark.

On the Mohs hardness scale, a person's fingernail measures 2.5. This means that gallium, indium, and thallium could all be scratched by your fingernail.

Reactivity

As with their physical properties, the chemical properties of the group 13 elements vary widely. Boron is much less reactive compared to the other members of the group. This means that it does not combine with other elements as readily. By contrast, the other group 13 elements are fairly reactive.

Aluminum is a very reactive metal. It reacts very easily with nonmetal elements, especially oxygen. When exposed to oxygen in the air, aluminum immediately reacts with the oxygen. That reaction forms a very thin coating of aluminum oxide (Al_2O_3) on the surface. This coating protects the aluminum from further reaction with oxygen. Aluminum also reacts with both acids and alkalis. An alkali is a chemical with properties that are opposite of those of an acid. Both acids and alkalis dissolve the protective aluminum oxide coating.

Gallium combines with most nonmetals when heated to high temperatures. Like aluminum, it also reacts with both acids and alkalis. Indium metal dissolves in acids but does not react with alkalis. Like gallium, it reacts with oxygen only at higher temperatures. Thallium reacts with acids and with the oxygen in the air. When exposed to air, a coating called thallium oxide (Tl_2O) forms on thallium's surface.

Chapter Four
Boron Group Compounds

A compound is formed when two or more elements bond together. There are millions of different compounds. The traits of a compound are often very different than the traits of the individual elements from which it was made. For example, by itself, aluminum is a soft silvery metal. However, aluminum combines with sulfur (S) and oxygen to form aluminum sulfate, a compound that has clear, colorless, brittle crystals.

Aluminum and boron both have many useful compounds. Aluminum is commercially the most useful of the group 13 elements. Aluminum compounds, such as aluminum sulfate, are used for paper treatment and water purification.

Aluminum is also combined with other metals to make alloys. Alloys are made by melting and then mixing two or more metals. The resulting mixture has properties that are different from the properties of the individual metals. Things like railroad cars, automobiles, and aircraft have many parts made from aluminum alloys. Aluminum alloys are strong and lightweight, and they are ideal in situations in which mobility and energy conservation are important.

Boron compounds are used in agriculture and to make glass and detergents. Boron's most important compounds are borax (sodium borate), boric oxide, and boric acid. Boric oxide is used in making

This boron mine located in California's Mojave Desert is the largest boron mine in the world.

heat-resistant glass found in cookware and scientific glassware. Boric acid is used as an insecticide, especially for ants and cockroaches.

Indium Compounds and Their Uses

Indium is primarily used to make alloys. It takes only a small amount of indium in an alloy to make a big difference. For this reason, indium has been referred to as the "metal vitamin." A very small amount of indium is sometimes added to platinum (Pt) or gold (Au) alloys to make them much harder. Alloys like these are used in electrical devices and in dental materials.

About 15 percent of the indium produced is used to make indium semiconductor compounds. A semiconductor is a solid material that has conductivity between that of a conductor and an insulator (a material that prevents the passage of heat or electricity). Indium compounds that are used to make semiconductors include the following: a combination of indium and antimony (Sb) called indium antimonide (InSb), a combination of indium and phosphorus (P) called indium phosphide (InP), and a combination of indium and nitrogen (N) called indium nitride (InN). Indium semiconductors are mainly used in high-speed transistors, infrared detectors, and photovoltaic devices.

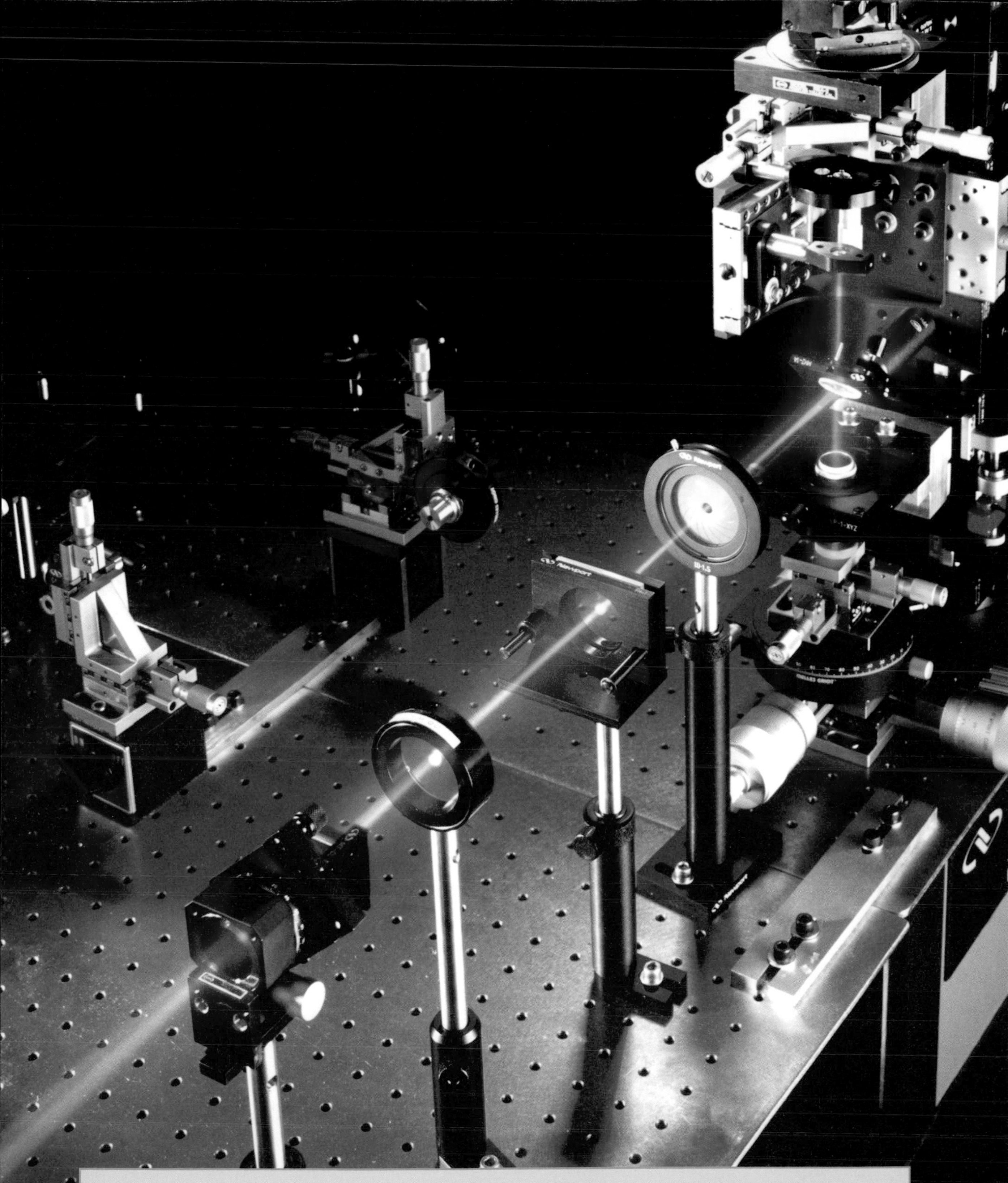

A laser can be used to measure the properties of a semiconductor thin film. This semiconductor thin film is a combination of gallium, indium, and phosphorus (GaInP).

High-speed transistors are involved in amplifying small electrical signals and in processing digital information. Infrared detectors are devices that respond to infrared radiation, a type of invisible light. These devices are used for such purposes as detecting fires and detecting overheating in machinery, vehicles, planes, and people. They also make it possible to view things in the dark. Humans and other mammals give off infrared radiation, sometimes called thermal light, and this can be detected with indium semiconductors. Photovoltaic cells, or solar cells, are used to change sunlight into electrical current.

In 1997, a solar conversion system was developed that uses a compound of copper (Cu) , indium, and selenium (Se), called copper indium diselinide ($CuInSe_2$), in its solar cells. This system is able to produce one million watts of electrical energy from sunlight. This is about the amount of electricity that it takes to operate a large office building. Many scientists feel that, in the future, solar cells will replace oil, coal, and natural gas as a source of energy.

Indium's very low melting point, 313.9°F (156.6°C), and its ability to conduct electricity make it a useful component in solders and low-melting alloys. When it is pure, indium sticks very tightly to itself and to other metals. This property makes indium very useful as a solder. A solder is a material that joins two metals together. About 12 percent of indium consumption is used to make low-melting alloys and solders. In these solders, the indium sometimes acts to reduce the melting point, sometimes to strengthen, and other times to prevent the solder from breaking down too easily.

Most indium is used to make coatings. Some aircraft parts are coated with indium-containing alloys. The indium in these alloys prevents the parts from wearing out or reacting with the oxygen in the air. Indium is also used to make thin film coatings. These are used to make electronic devices, such as liquid crystal displays (LCDs). The compound indium tin

Indium metal, like most metal, is ductile. This means that it can be drawn into wire.

oxide (ITO) is used to make LCDs and accounts for about half of indium's annual consumption. A liquid crystal is a liquid that becomes clear or cloudy, depending on the temperature or applied voltage. LCDs allow characters, such as the numbers or letters in a calculator, to be displayed. They are used in electronic devices like laptops, flat-panel televisions, flat-panel monitors, cell phones, PDAs, digital cameras, clocks, watches, and GPS receivers.

The health effects of indium compounds are interesting. If indium compounds are consumed through the mouth, they are relatively harmless. However, if they are injected through the skin, they are very poisonous.

Gallium Compounds and Their Uses

An important use for gallium compounds is as semiconductors. Two examples of gallium compounds that are used as semiconductors are gallium nitride (a combination of gallium and nitrogen) and gallium arsenide (a combination of gallium and arsenic [As]). About 98 percent of the gallium that is produced is used to make these two compounds.

These semiconductors are mainly used in light-emitting diodes (LEDs). An LED is a device that gives off visible light or infrared radiation when an electric current passes through it. LEDs are used as indicator lights on electronic devices, in flashlights, and in area lighting. The color of the light produced depends on what the semiconductor is made of. For example, the compounds gallium nitride and gallium phosphide produce green light. UV-LEDs are also used for sterilization of water and to disinfect devices. They are also used as grow lights to help plants carry out photosynthesis. Some police light bars use LED technology. Gallium arsenide is also used to make lasers. An electric current is passed into a piece of gallium arsenide, and a beam of laser light is produced. Gallium arsenide lasers are used for many things, including compact disc (CD) players.

Gallium semiconductors are used in integrated circuits as well. Integrated circuits are also known as IC, microcircuits, microchips, silicon chips, or simply as chips. They are miniaturized electronic circuits, made mainly of semiconductor devices that have been manufactured into the surface of another semiconductor material. Integrated circuits are found in almost all of the electronic equipment that we use today. They have revolutionized the world of electronics by making electronic devices less expensive and more durable.

Other gallium compounds are also useful. Gallium readily alloys with most metals. An alloy of gallium, indium, and tin is widely used in medical thermometers. This alloy, called galinstan, has a freezing point

The huge Sky Screen in Beijing, China, is covered entirely with LED modules.

of -4°F (-20°C). Galinstan in thermometers is replacing mercury, an element that is toxic to both humans and animals. Gallium is also used in nuclear weapon pits, the core of the weapon. Gallium is alloyed to the plutonium (Pu) explosive used in nuclear weapon pits. The gallium prevents the plutonium from becoming brittle as it cools, so it can be machined into the needed shape during manufacture. Gallium compounds, such as gallium arsenide (GaAs) and copper indium gallium selenium sulfide ($Cu(In,Ga)(Se,S)_2$), are also used in photovoltaic cells. Copper indium gallium sulfide has been found to be a more efficient alternative to crystalline silicon in solar panels. Solar panels contain solar cells that convert the sun's radiation into electricity.

Gallium and its compounds can be hazardous to the health of humans and animals. They produce skin rashes and a decrease in the production of blood cells. Therefore, gallium and its compounds should always be handled with care. A sign of exposure to gallium and its compounds is a characteristic metallic taste in the mouth.

Thallium Compounds and Their Uses

Even though thallium is a pretty uncommon element, some of its compounds have very important applications. About 60 to 70 percent of the thallium that is produced is used in the electronics industry. Thallium sulfide (Tl_2S) is sometimes used in photocells. Photocells convert light into electrical energy. Thallium sulfide does not conduct electricity very well in visible light. When exposed to infrared light, however, the compound conducts electricity very well. Therefore, special photocells can be built that are used to detect infrared radiation. Thallium compounds, such as thallium oxide (Tl_2O_3), are also used in making superconducting materials. Superconductors have no resistance to the flow of electricity. Once electricity begins to flow in a superconductor, it flows forever. Superconducting materials that contain thallium are used for such things as magnetic resonance imaging (a medical technique used for diagnosis that creates thin-section images of a person's body parts, such as the heart), magnetic propulsion (a type of movement that uses magnetic forces to lift, propel, and guide an object, such as a train), and electric power generation.

Thallium is highly toxic. For this reason, one of its compounds, thallium sulfate (Tl_2SO_4), was once used in rat poisons (rodenticide) and ant killers (insecticides). Because thallium sulfate is colorless and odorless, rats and ants are not aware that the compound is present. Thallium sulfate passes quickly through the skin. Once inside the body, it causes

Superconducting materials that contain thallium are used in MRI machines in order to create thin-section images of organs such as the heart.

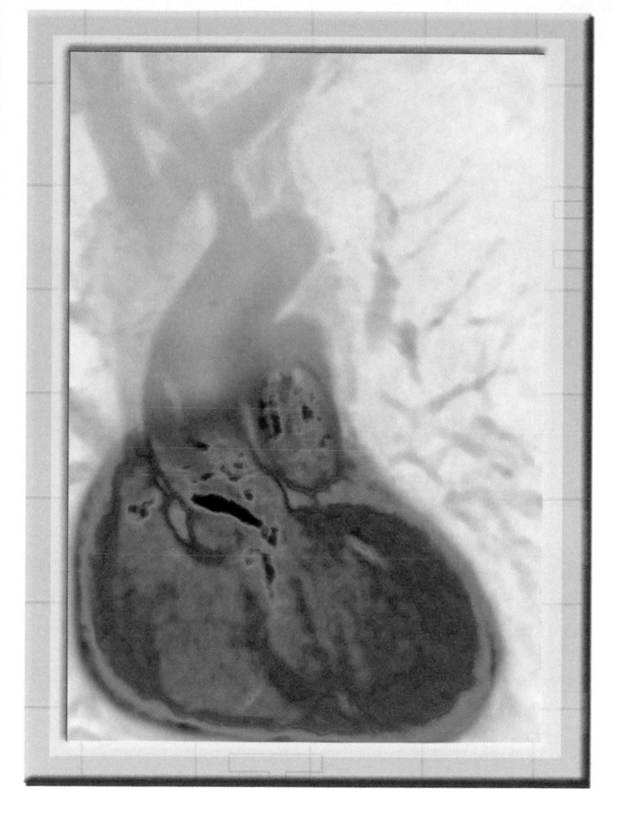

death. Unfortunately, thallium sulfate has the same effect on humans. There have been many cases of accidental poisonings, especially of children. Therefore, the use of thallium compounds in rodenticides and insecticides has been reduced or eliminated in many countries. The United States has had a ban on thallium sulfate poisons since 1975.

Chapter Five
The Boron Group and You

Boron is an essential element for plants. Essential elements are those that are needed for an organism to survive and thrive. Because boron is essential for plants, people regularly consume about 0.07 ounces (2 grams) of boron a day in the foods that they eat. Foods that contain the most boron are fruits (especially apples, peaches, grapes, pears, and cherries) and vegetables (especially broccoli, onions, and cabbage). In the human body, small amounts of boron are found in teeth and in bones. Too much boron can sicken a person, but it would take quite a bit of boron to make a person ill (about 0.17 oz [5 g]) and even more to put a person's life in danger (about 0.7 oz [20 g]).

Aluminum is not known to be essential for any living thing. However, because there is so much aluminum in the soil, all plants absorb it. Some of the crops that we eat contain aluminum, such as spinach, oats, lettuce, onions, and potatoes. People also consume aluminum from the tea that they drink. Aluminum is no longer considered dangerous at the levels normally consumed by people. However, some physicians believe that a large buildup of aluminum in the body could cause health problems, such as lung disease and Alzheimer's.

Gallium and indium have no biological roles, but they are known to stimulate metabolism. Metabolism is the series of chemical reactions

All of the group 13 elements (except indium) are readily absorbed by plants from the soil in which they grow. When we eat vegetables, such as the spinach seen here, we also consume these elements.

taking place in a living organism. Plants absorb a small amount of gallium from the soil in which they grow. Therefore, the vegetables that people eat can contain a very small amount of gallium. Gallium is not toxic and presents few health risks.

However, indium is toxic if more than a few milligrams are consumed. It can affect the heart, the liver, and the kidneys. There is also no biological role for thallium. However, about trace amounts of thallium are consumed by the average person each day as part of his or her diet. The thallium builds up in a person's body over time, concentrating in the kidneys and liver. Thallium is a highly toxic metal. It can even be absorbed through the skin.

Medical Roles

One of aluminum's medical uses is in an over-the-counter treatment to help stop bleeding. It is sold in the form of styptic (able to stop bleeding) pencils that contain aluminum sulfate and that can be used on minor cuts, such as those experienced from shaving. Aluminum chloride or aluminum chlorhydrate is the active ingredient in most antiperspirants. The other group 13 elements have isotopes that have useful medical roles.

Understanding Isotopes

Isotopes are atoms of an element that have different numbers of neutrons. In nature, almost all elements exist as a mixture of two or more isotopes. For example, there are two naturally occurring isotopes of gallium: gallium-69 and gallium-71. These numbers are the mass numbers of the isotopes. The mass number is the number of protons plus neutrons in the isotope's nucleus. Gallium atoms always have thirty-one protons (that's what makes it gallium), so gallium-69 and gallium-71 have thirty-eight and forty neutrons, respectively.

About a dozen artificial radioactive isotopes of gallium are known. A radioactive isotope breaks apart and gives off some form of radiation. Artificial isotopes can be made by exposing natural isotopes to extremely high-energy radiation, such as that from a nuclear reactor. One radioactive isotope of gallium, gallium-67, has been used in medicine for a long time. This isotope seeks out cancer cells in the body. The radiation that the isotope gives off can then be detected and lead doctors to the location of cancer. Gallium-67 has been used to look for cancer in the spleen, the liver, the bowels, the kidneys, the breasts, and the bones.

There are two naturally occurring isotopes of indium: indium-113 and indium-115. However, a number of artificial radioactive isotopes of indium also exist. Two of indium's isotopes are used in medicine.

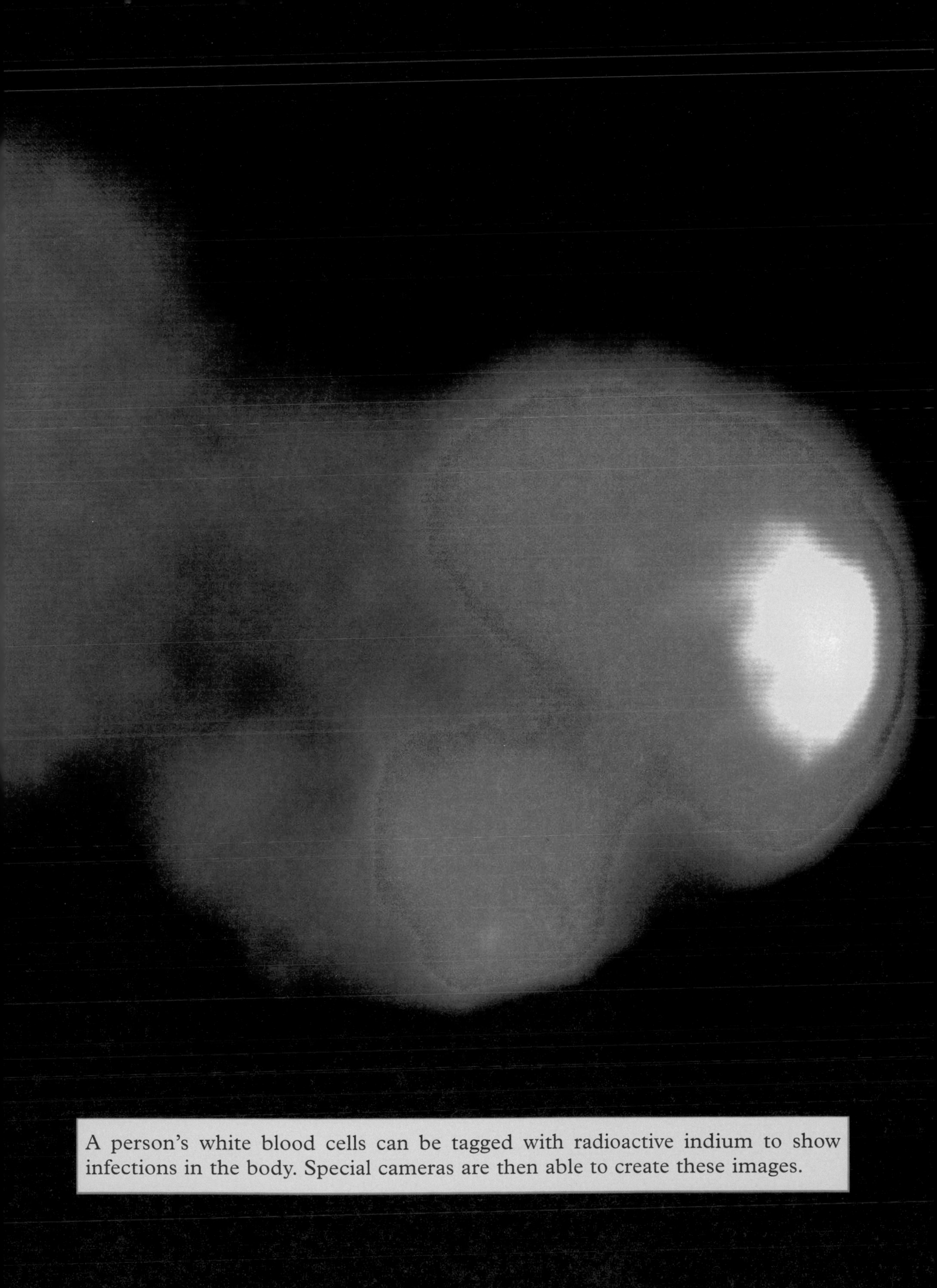

A person's white blood cells can be tagged with radioactive indium to show infections in the body. Special cameras are then able to create these images.

Indium-113 is used to examine the spleen, the brain, the liver, the pulmonary system (the lungs), and the circulatory system (the heart and blood). Indium-111 is used to search for tumors, abscesses (puss-filled cavities), internal bleeding, and infections. It is also used to study the gastric (stomach) and blood systems. The isotope is injected into a person's bloodstream, and the radiation it gives off can be detected by a camera or other device.

Thallium has two naturally occurring isotopes, thallium-203 and thallium-205. About two dozen radioactive isotopes of thallium have also been made. One of these, thallium-201, is used to study the flow of blood in the body. It shows how well the heart is working. Many times, the

Molten indium and gallium are able to coat nonmetallic materials. This process is called wetting.

isotope is used in stress tests. Thallium-201 is injected into a patient's bloodstream. The person exercises and then lies down. A camera that is able to record the radiation given off by the isotope is then passed over the patient's body. This can tell doctors if the person's heart is working properly or not.

Glass Wetting

Indium and gallium have the ability to coat nonmetallic materials when the molten metals are gently rubbed against the surface of the nonmetal. In this way, indium and gallium are able to "wet" almost any surface. Instead of beading up, they spread out over the surface. This process makes the metals good for making mirrors. Indium deposited on glass produces a mirror as good in quality as that of silver but more resistant to corrosion. Gallium coating can be used to make objects electrically conductive.

The Periodic Table of Elements

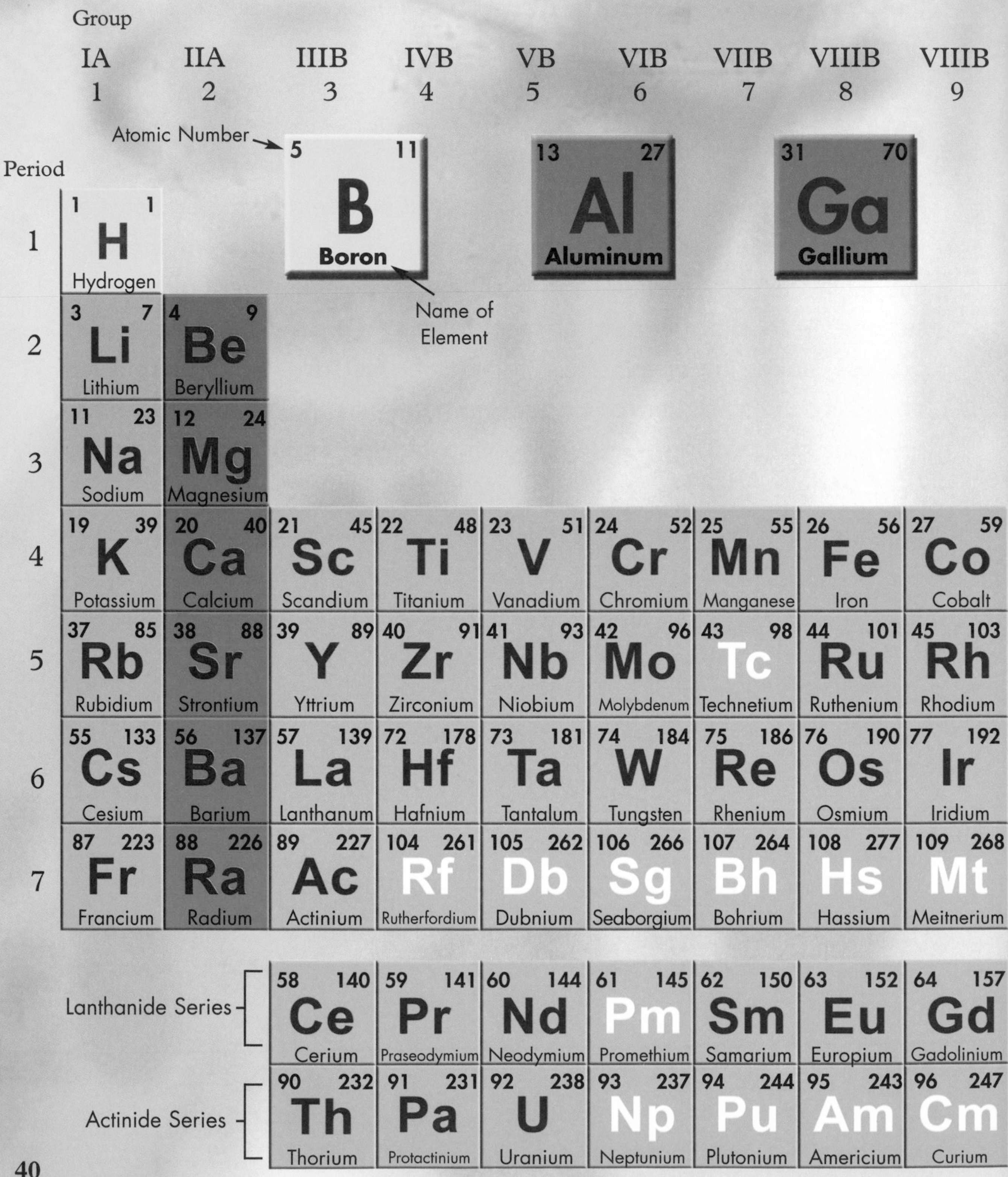

Period	IA 1	IIA 2	IIIB 3	IVB 4	VB 5	VIB 6	VIIB 7	VIIIB 8	VIIIB 9
1	1 H Hydrogen 1								
2	3 Li Lithium 7	4 Be Beryllium 9							
3	11 Na Sodium 23	12 Mg Magnesium 24							
4	19 K Potassium 39	20 Ca Calcium 40	21 Sc Scandium 45	22 Ti Titanium 48	23 V Vanadium 51	24 Cr Chromium 52	25 Mn Manganese 55	26 Fe Iron 56	27 Co Cobalt 59
5	37 Rb Rubidium 85	38 Sr Strontium 88	39 Y Yttrium 89	40 Zr Zirconium 91	41 Nb Niobium 93	42 Mo Molybdenum 96	43 Tc Technetium 98	44 Ru Ruthenium 101	45 Rh Rhodium 103
6	55 Cs Cesium 133	56 Ba Barium 137	57 La Lanthanum 139	72 Hf Hafnium 178	73 Ta Tantalum 181	74 W Tungsten 184	75 Re Rhenium 186	76 Os Osmium 190	77 Ir Iridium 192
7	87 Fr Francium 223	88 Ra Radium 226	89 Ac Actinium 227	104 Rf Rutherfordium 261	105 Db Dubnium 262	106 Sg Seaborgium 266	107 Bh Bohrium 264	108 Hs Hassium 277	109 Mt Meitnerium 268

Series							
Lanthanide Series	58 Ce Cerium 140	59 Pr Praseodymium 141	60 Nd Neodymium 144	61 Pm Promethium 145	62 Sm Samarium 150	63 Eu Europium 152	64 Gd Gadolinium 157
Actinide Series	90 Th Thorium 232	91 Pa Protactinium 231	92 U Uranium 238	93 Np Neptunium 237	94 Pu Plutonium 244	95 Am Americium 243	96 Cm Curium 247

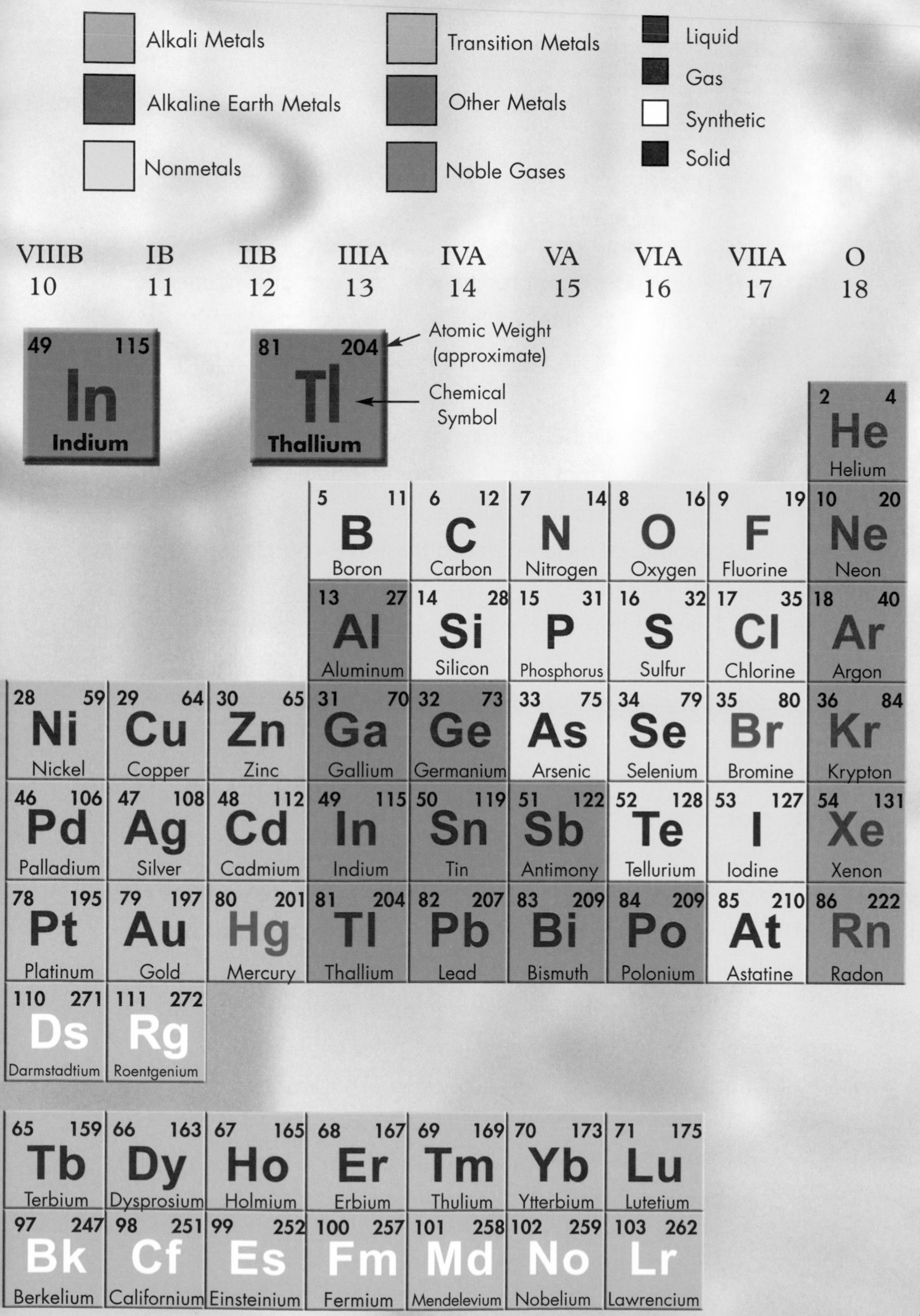

Alkali Metals
Alkaline Earth Metals
Nonmetals
Transition Metals
Other Metals
Noble Gases
Liquid
Gas
Synthetic
Solid
VIIIB 10
IB 11
IIB 12
IIIA 13
IVA 14
VA 15
VIA 16
VIIA 17
O 18
49 115 In Indium
81 204 Tl Thallium
Atomic Weight (approximate)
Chemical Symbol
2 4 He Helium
5 11 B Boron
6 12 C Carbon
7 14 N Nitrogen
8 16 O Oxygen
9 19 F Fluorine
10 20 Ne Neon
13 27 Al Aluminum
14 28 Si Silicon
15 31 P Phosphorus
16 32 S Sulfur
17 35 Cl Chlorine
18 40 Ar Argon
28 59 Ni Nickel
29 64 Cu Copper
30 65 Zn Zinc
31 70 Ga Gallium
32 73 Ge Germanium
33 75 As Arsenic
34 79 Se Selenium
35 80 Br Bromine
36 84 Kr Krypton
46 106 Pd Palladium
47 108 Ag Silver
48 112 Cd Cadmium
49 115 In Indium
50 119 Sn Tin
51 122 Sb Antimony
52 128 Te Tellurium
53 127 I Iodine
54 131 Xe Xenon
78 195 Pt Platinum
79 197 Au Gold
80 201 Hg Mercury
81 204 Tl Thallium
82 207 Pb Lead
83 209 Bi Bismuth
84 209 Po Polonium
85 210 At Astatine
86 222 Rn Radon
110 271 Ds Darmstadtium
111 272 Rg Roentgenium
65 159 Tb Terbium
66 163 Dy Dysprosium
67 165 Ho Holmium
68 167 Er Erbium
69 169 Tm Thulium
70 173 Yb Ytterbium
71 175 Lu Lutetium
97 247 Bk Berkelium
98 251 Cf Californium
99 252 Es Einsteinium
100 257 Fm Fermium
101 258 Md Mendelevium
102 259 No Nobelium
103 262 Lr Lawrencium

Glossary

absolute zero The lowest possible temperature, about -459°F (-273°C).

alkali (base) A substance that reacts with an acid and neutralizes it.

alloy A mixture of two or more metals.

atom The smallest part of an element having the chemical properties of that element.

atomic number The number of protons in the nucleus of an atom.

bond An attractive force that links two atoms together.

conductivity The ability to transmit electricity.

electron A subatomic particle having a negative charge and a very small mass.

element A substance that cannot be separated into other substances by a chemical reaction.

group A set of similar elements found in the same column of the periodic table.

isotope Two or more forms of an element that differ from each other by their mass numbers.

mass The property of matter that gives matter its weight.

matter What objects are made of. Matter takes up space and has mass.

neutron A subatomic particle with no electrical charge that has a mass approximately equal to that of a proton.

ore A material found in the earth from which a useful substance (usually a metal) can be extracted.

proton A subatomic particle with a positive electrical charge and a mass approximately equal to that of a neutron.

smelting The process of melting an ore to obtain a metal from it.

volume The amount of space that something occupies.

For More Information

AIM Specialty Materials USA
25 Kenney Drive
Cranston, RI 02920
(401) 463-5605
Web site: http://www.aimspecialty.com
AIM specializes in alloys of indium, tin, lead, bismuth, cadmium, gold, and gallium.

Aluminum Company of America
Alcoa Corporate Center
201 Isabella Street
Pittsburgh, PA 15212-5858
(412) 553-4545
Web site: http://www.alcoa.com/global/en/home.asp
The Aluminum Company of America produces primary aluminum, fabricated aluminum, and alumina combined.

Indium Corporation
P.O.B. 269
Utica, NY 13503
(800) 4-INDIUM
Web site: http://www.indium.com
The Indium Corporation extracts, refines, purifies, and fabricates indium metal products.

Mineral Information Institute
505 Violet Street

Golden, CO 80401
(303) 277-9190
Web site: http://www.mii.org
The Mineral Information Institute provides students and teachers with materials that help them understand that mineral and energy resources are essential to society.

Teck Cominco Limited
600 - 200 Burrard Street
Vancouver, BC V6C 3L9
Canada
(604) 687-1117
Web site: http://www.teckcominco.com
Teck Cominco Limited is a mining company that produces copper, zinc, metallurgical coal, and specialty metals.

Web Sites

Due to the changing nature of Internet links, Rosen Publishing has developed an online list of Web sites related to the subject of this book. This site is updated regularly. Please use this link to access the list:

http://www.rosenlinks.com/uept/tbe

For Further Reading

Adair, Rick. *Boron*. New York, NY: Rosen Publishing, 2007.

Baldwin, Carol. *Compounds, Mixtures, and Solutions* (Material Matters/Freestyle Express). Chicago, IL: Raintree, 2004.

Beatty, Richard. *Boron*. New York, NY: Benchmark Books, 2005.

Hasan, Heather. *Aluminum*. New York, NY: Rosen Publishing, 2007.

Knapp, Brian. *Elements: Aluminum*. New York, NY: Grolier Educational, 2002.

Newton, David E. *The Chemical Elements*. New York, NY: Franklin Watts, 1994.

Oxlade, Chris. *Elements and Compounds*. Chicago, IL: Heinemann, 2002.

Oxlade, Chris. *Metals*. Chicago, IL: Heinemann, 2002.

Sacks, Oliver. *Uncle Tungsten: Memories of a Chemical Boyhood*. New York, NY: Vintage Books, 2002.

Saunders, Nigel. *Aluminum and the Elements of Group 13* (The Periodic Table). Chicago, IL: Heinemann, 2004.

Stille, Darlene. *Chemical Change: From Fireworks to Rust*. Minneapolis, MN: Compass Point Books, 2005.

Stwertka, Albert. *A Guide to Elements*. New York, NY: Oxford University Press, 2002.

Tocci, Salvatore. *The Periodic Table*. New York, NY: Children's Press, 2004.

Bibliography

Chang, Raymond. *Chemistry*. 9th ed. New York, NY: McGraw-Hill, 2006.

Ebbing, Darrell D., and Steven D. Gammon. *General Chemistry*. 9th ed. New York, NY: Houghton Mifflin Company, 2009.

Emsley, John. *Nature's Building Blocks: An A–Z Guide to the Elements*. New York, NY: Oxford University Press, 2003.

Schwarz-Schampera, Ulrich, and Peter M. Herzig. *Indium*. New York, NY: Springer, 2002.

Stwertka, Albert. *A Guide to the Elements*. 2nd ed. New York, NY: Oxford University Press, 2002.

Index

About the Author

Heather Elizabeth Hasan graduated from college summa cum laude with a dual major in biochemistry and chemistry. She has written numerous books for Rosen Publishing, including *Understanding the Elements of the Periodic Table: Iron*, *Understanding the Elements of the Periodic Table: Helium*, *Understanding the Elements of the Periodic Table: Fluorine*, *Understanding the Elements of the Periodic Table: Nitrogen*, and *Understanding the Elements of the Periodic Table: Manganese*. She currently lives in East Stroudsburg, Pennsylvania, with her husband, Omar; their sons, Samuel and Matthew; and their daughter, Sarah.

Photo Credits

Cover, pp. 1, 5, 13, 15, 40–41 by Tahara Anderson; p. 7 © Roger Harris/Photo Researchers, Inc.; p. 8 © Michael Maloney/San Francisco Chronicle/Corbis; p. 9 © SSPL/The Image Works; p. 18 © Richard Treptow/Photo Researchers, Inc.; p. 20 © Lester V. Bergman/Corbis; p. 22 © Joel Arem/Photo Researchers, Inc.; p. 26 © Jim West/The Image Works; p. 27 © Science Source/Photo Researchers, Inc.; pp. 29, 38 Wikimedia Commons; p. 31 © Bruce Connolly/Corbis; p. 33 © James Cavallini/Photo Researchers, Inc.; p. 35 Shutterstock.com; p. 37 cmsphoto/Newscom.

Designer: Tahara Anderson; Editor: Nicholas Croce;
Photo Researcher: Cindy Reiman